# HOW TO GROW
# MARIJUANA

:The Beginners guide to growing,caring and
harvesting cannabis from seed

*Larry Pat*

1

# TABLE OF CONTENT

# INTRODUCTION

*Pot, hashish, pot, cannabis…* You most likely know all about the different weed names. **However, how does one grow the thing at home? How would you get your seeds to grow, enter the vegetative state, and eventually produce buds?**

**How will you even tell when your marijuana plant is ready to harvest ?**

**Furthermore, how would you dry and cure the harvested stuff?**

Be at ease, enthusiastic first-time growers (perhaps even for the second or third time?). we take care of you! We've made sense of everything in straightforward language so you can begin immediately. Furthermore, no, you don't need to be a specialist in the game, yet our unrivaled aide takes you by the hand and leads you on. :)

So, let's get started!

# GETTING STARTED

**First thing first**- understanding what you want will set you up for the game.

**Grow medium**: natural soil isn't your main decision

**Grow Light**: yields are affected by brightness in water

**Air**: The right pH (in water) is important. think about outside air (with a slight breeze)

**Temperature**: nutrients: not too hot nor too cold like different plants, weed plants need to eat

These focuses are only a synopsis of the significant things you'll have to develop pot — we'll examine it exhaustively in the resulting segments. Stay nearby.

# STEP 1: DECIDE WHERE YOU ARE GOING TO GROW YOUR CANNABIS.

*Should you grow it indoors or outdoors?*

Indeed, each accompanies its advantages and disadvantages. Developing a pot outside is less expensive.

At the point when you develop outside, you don't need to give a large portion of the provisions as the earth's life force and direct daylight will deal with that.

Notwithstanding, your plants are probably going to be taken, pollinated, pervaded by bugs, eaten by deer, and so on.

Protection isn't additionally insured — judgemental neighbors around could start the blame-shifting game. You get the point. Establishing weed inside, in the interim, could accompany a greater cost tag, in all honesty. You'll have to pick an ideal developed space, introduce a dependable lighting framework, control temperature/mugginess, and so on.

However, the advantages far offset the expenses (assuming you understand what you're doing). You have absolute command over your pot plants — you can "tell" them to blossom by changing the lighting cycle, in addition to other things.

Simply guarantee you're giving all that your plants need to develop — supplements, water, temperature, stickiness, and so on.

Any other way, you could think twice about rich development.

Probably the prescribed procedures you would rather not disregard incorporate the accompanying:

**Begin little**: particularly on the off chance that you're a first-time producer.

**Keep it clean**: from the seed to the harvest

**Keep it light-close:** so light doesn't get away from through certain outlets.

Maintain the appropriate airflow, humidity, and temperature.

Ensure you're ready to cover that power bill, however, it can soar, particularly assuming that you're a huge indoor producer.

Guarantee the arrangement costs are acceptable for you, as well.

## STEP 2: CHOOSE A CANNABIS GROWING MEDIUM

Try not to avoid giving a shot at different developed mediums other than soil. We should list and make sense of it exhaustively.

## SOIL

It's given this is the most famous developing medium, presumably because it's promptly accessible and has a few prepared supplements in it. Soil-developed weed is likewise more delicious and more fragrant. Furthermore, no, the developing system isn't testing - simply stick your (liked) strain in the dirt and go.

Period. By the way, while there are different sorts of soil, excellent fertilized soil will make for a speedy, basic method for the beginning. You should add to it a few exceptionally made supplements, as well, for flourishing pot development.

Then again, you can create your super soil. If you pick to purchase soil, we suggest you go for top brands, basically those with an ideal soil blend. They ought to contain such things as worm castings, horse feed feast, humic corrosiveness, and so forth.

## SOILLESS

**Want to eliminate soil?**

Prepare to experience increased yields and faster growth. Additionally, advantageously, developing weed in soilless mediums is pretty much in developing the thing in the dirt. Just you'll have to give the supplements without any preparation, not at all like with soil, which brags supplements on its own. Instances of soilless mediums incorporate coco coir, vermiculite, perlite, and so forth.

## HYDROPONIC GROW

No, aquaculture development isn't muddled. It's just about as straightforward as picking your arrangement (aquaculture framework), getting your supplements, and growing their seeds. Furthermore, you'll be compensated with inconceivably quick development, astounding yields, and unparalleled weed power.

Something you need, correct? Be careful not to fall for incorrect information, which could jeopardize everything. Yet, with the right arrangement and upkeep, have confidence a plentiful gathering is coming up for you.

# STEP 3: CHOOSE NUTRIENT

Like most plants, pot plants need to "eat." You just need to understand what supplements to use for your developing medium: soil, soilless, and aquaculture. It's simply that different weed-developing mediums have various supplements intended for them - and involving some unacceptable supplements for your picked developed medium could think twice about rich development.

Utilizing cannabis supplements is super simple, given you're sticking to their supplement and taking care of diagrams
(adhering to guidelines exactly). Still…. Except if your plants will be ready to assimilate the supplements employing the plant roots, taking care of them is close to sitting idle. Try to keep up with the right PH level of the root climate, as well, particularly on the off chance that you're utilizing fluid supplements.
And it's easy. Change the water's PH before watering your plants; a uniquely planned PH unit will prove to be useful here. Not all that quick: Different PH levels are required for different grow mediums.

Assuming you will utilize soil, guarantee the PH is wavering in the 6.0 to 7.0 territory. For aqua farming, the PH ought to fall somewhere in the range of 5.5 and 6.5. Try to adhere to these "rules" strictly. That is if you want your cannabis plants to produce more yields and have denser buds.

## STEP 4: CHOOSE AN EASY TO GROW CANNABIS STRAIN FOR BEGINNERS

**Is it safe to say that you are a newcomer?** Just relax - there's something planned explicitly for you. **Have you ever known about auto-flowering weed seeds?** We'll say why. These are weed seeds that needn't bother with an exceptional lighting cycle to invigorate the maturing stage.

This has two advantages. As a matter of some importance, you will not need to burn through cash on costly light clocks. Second, you won't be late or early when changing the light cycle, so you'll always get good yields.

**Who wouldn't desire that?**

Autoflowering strains likewise brag a more limited blossoming time than photoperiod plants, and that implies that upkeep time is diminished on your end. Another choice you need to get?

**Feminized seeds:**

Unlike regular seeds, which produce both male and female plants, these marijuana strains are guaranteed to flower because they produce female plants. As a result, you can avoid the guesswork associated with the entire budding process.

We strongly suggest that you select cannabis seeds that possess both of these characteristics: feminized and auto-flowering. Assuming you're pondering where you will get this from, ILGM has got you covered. The respectable seed bank sells an entire heap of novice strains moderately. In addition, ILGM allows you to channel the choices in light of key measurements, including THC and CBD levels. It's essential to make reference to that ILGM has a stockroom in the US, as well, for facilitated delivery.

Some ILGM beginners STRAIN include :
*Super Skunk Feminized*

*Gold Leaf Autoflower*
*Harsh Diesel Autoflower*
*Blueberry Autoflower*
*Aurora Borealis Autoflower*
This seed store sells blend packs which are only a bundle of various strains (up to three) at a limited cost. Ideal for those on a tight budget. Models incorporate the accompanying
*Beginners mix*
*Auto flower Classics mix*
*Auto flower CBD mix*
*Super Mix*
*Auto flower Snack Mix*

## SET UP YOUR LIGHTING

Fledglings frequently miss the point, which prompts lower yields... in all actuality, weed plants need various measures of light in the vegetative and blossoming stages. Presently, clearly, that is just valid for indoor cultivators - as Mother Earth will deal with that for open-air producers. So, exactly how much light do you require? You'll have to give 18 hours of light (a day) during the vegetative stage prior to changing to the 12-hour lighting cycle to set off the blooming stage.

Simply ensure your lighting framework is sufficiently tight. Light will in general hole during the dim periods, undermining your plant's development stages. These frequently revert to a different stage or produce male buds, for instance.

Here are the various kinds of weed development lights you need to utilize.

**Stowed away develop lights**

Focused energy release (Stowed away) lights create all the more light per unit of power utilized. They aren't generally so proficient as the Drove lighting apparatuses, however, yet they are the business standard and are the go-to choice by numerous cultivators. In any case, here's the inquiry... What Stowed away develop lights would it be advisable for you to use for your plants' different development stages? Indeed, there are two significant kinds of **Concealed lights**: sodium at high pressure and metal halide (MT).

**MT lights**: Due to their bluish-white light, these are extremely helpful during the vegetative stage.

**HPS lights:** These ones prove to be useful in the blossoming stage. They radiate rosy orange light. For your HID lighting setups, be sure to provide ballasts and reflectors—one for each bulb.

**NOTE**: Concealed bulbs produce a lot of intensity. Thus, you'll have to mount your lights in air-cooled reflector hoods, particularly on the off chance that you're hoping to fill in a little space with less ventilation. Such things as ducting and debilitating fans ought to prove to be useful here.

## FLUORESCENT GROW LIGHTS

High-yield (bright light bulbs) are very normal among limited-scope weed producers. They just cost less to set up: weight and hood are incorporated with every bulb, not at all like with the Concealed lights. Additionally, they do not require cooling systems because they produce less heat than HIDs do, saving you money. However, here is a bummer. Glaring lights aren't as proficient, creating some 30% less light (per unit of power utilized) than the Concealed light installations.

## LED GROW LIGHTS

While they can accompany a far greater cost tag, Light Transmitting Diode (Drove) bulbs brag an entire heap of advantages. Make the least amount of heat, last a long time, and use less electricity. The best designs also produce a wider spectrum of light.

Do you know what it means? Great yields.

## INDUCTION GROW LIGHTS

This is one more sort of lighting framework you can use to develop weed inside — for the right reasons. Steady light range Durable bulbs Low vacillations in light force

The two fundamental sorts of enlistment development lights are plasma and attractive.

Attractive acceptance lighting upholds both vegetative and blossoming stages — because of their better light entrance.

Plasma enlistment development lights, in the meantime, are deficient with regard to supporting lush cannabis development. They don't get enough of the spectrum of light to allow for healthy growth.

Thus, it's logical your plants will be hindered or shrink out and out assuming you will utilize just this sort of lighting in your developing region. This likewise presumably makes sense of why plasma enlistment lighting isn't suggested by master producers.

## STEP 6: DECIDE HOW YOU WANT TO MONITOR AND CONTROL THE CLIMATE.

Outside natural circumstances can think twice about indoor climate, harming solid plant development. For this reason, you need to screen and control your indoor natural circumstances in a manner - from dampness to wind stream and temperature environments.

In the event that your develop space is a wet cellar, for example, you ought to run a radiator (or dehumidifier) to settle the temperature. In the meantime, in the event that your developing region is excessively hot, your smartest option is to introduce AC units (or utilize additional fans) to cool the thing.

Another trick is to turn your lights on at night when it's cold outside and off during the day when it's hot. Possibly you'll have the option to work on your plants when the lights are on — likely around evening time.

## STEP 7: PICK A GOOD CONTAINER FOR YOUR PLANTS

There are a lot of holders that make for a rich indoor development climate. Furthermore, the choices are boundless, from ceramic pots to texture compartments and customary plastic holders. However, you need to snatch appropriate develop holders, essentially founded on your weed needs - plant size, thickness, and so forth. In this way, guarantee these are checking the right boxes, particularly with regards to supporting flourishing underground root growth.

A portion of what to consider prior to taking out your wallet incorporates the accompanying.

**Nutrients**: Your grow container should help ensure the right temperature, soil PH, and other factors to ensure optimal nutrient absorption in order to prevent deficiencies.

**Oxygen**: Guarantee your holder is punctured fittingly to consider oxygen. A robust root system will be made possible by this. Drainage: Your compartment ought to have the option to hold water so your weed plants don't shrivel and pass on. Just it ought to be at moderate levels; over-the-top water can kill your plants because of root decay.

**Space:** Uncovers really do branch while they create. This could be compromised and the plant choked out by a smaller container. It doesn't make any difference to assume that you're hoping to sprout seeds or relocate the seedlings. Yet, ensure your develop holders can support a flourishing underground root growth by remembering the above factors.

## STEP 8: GROW YOUR FIRST MARIJUANA PLANT

Now is the time to get serious about it; grow your first marijuana plant by following these steps.
Sow your seeds
Water your plants
Give natural supplements
Screen light, temperature, air, and so forth.

Spot the distinction between vegetative and blooming development stage

Gather

**Develop Your Pot Seeds**

On the off chance that you ended up finding a companion that stays faithful to his obligations, you could have gotten a completely mature plant or a clone seedling. All things considered, you don't have to do any seed germination since it has proactively been finished for you.

Germination is the point at which a seedling sprouts from a seed interestingly. When the seed has been developed, it very well may be moved to the developing medium.

On the off chance that you are beginning without any preparation, you'll have two germination techniques you can utilize:

## GERMINATION TRAYS

These are ordinary plastic plates with little openings that can hold modest quantities of soil with a seed inside them. They can be watered effectively and productively.

A few plates have warming frameworks that add to their flexibility and straightforwardness. In other circumstances, you can simply use your lighting to apply heat. One way or the other, you ought to have a brief look at your most memorable seedlings in 3-7 days.

When the seedlings have grown you can take the entire 3D shape of soil in the plate and plant them where you anticipate developing the experienced blossom later down the line.

## THE PAPER TOWEL METHOD

This is a method for growing cannabis seeds. You might have already used the paper towel method when you were in fifth grade. In the event that you haven't, then you most likely currently suspect it's very easy to do.

Simply place several seeds in a soggy paper towel. Place the paper towel on a plate, fold it in half, and cover it all with a second plate. On the off chance that plates don't seem like what a genuine producer utilizes, then, at that point, you could possibly track down something to work with at your go-to home improvement shop.

Once the seedlings have sprouted, take them and plant them in the growing medium of your choice. Regularly monitor the germination process.

## VEGETATIVE STATE

I like saying that the vegetative stage is the piece of the cycle when you feel more 'on top of' your plants. This is the point at which you must be the most reasonable and notice how the plant is 'conversing with you'.

Your plant transitions from the seedling stage to the real plant at this point in the process. Although the growth of the leaves has begun, there is no flower to trim yet. You're checking out at a youthful plant by this point.

The plant will look miserable or solid and you must break down what this implies. Maybe she's asking you for pretty much water, a supplement portion change, or a temperature change. Getting to realize your plant is all important for the cycle and the Good times.

## FLOWERING STAGE

Developing your buds is possibly the justification for why you need to begin developing your plants. Developing further yielding plants is the intriguing part since it shows you the aftereffects of all the difficult work and care you've placed into your reap. Recall it is at this stage that you'll need to bring down your temperatures to 18-26 C.

The blossoming stage is described by two primary assignments you should achieve:

**1. Changing Your Light schedule**
In the event that you're developing your weed inside, you'll need to change your light timetable to 12 hours of all-out murkiness and 12 hours of light. Which ought to be not difficult to do in an indoor developing climate.

If you grow your plants outside, you won't be able to do this right away.

You'll need to make arrangements for it ahead of time from the germination stage. Plan out your development cycles so your plants arrive at their blossoming stage when the days get more limited, which ought to be at some point during pre-winter or fall.

## 2. Identifying Your Plant's Gender

Plant orientation is significant in light of the fact that it will have an effect on whether you are developing blossoms you can smoke or just weed plants for it.

Male plants do not produce flowers, whereas female plants do. Meaning, you can collect smokable blossoms from female weed plants.

Females are effectively recognizable on the grounds that they will begin developing pistils on the joints of the top branches. Pistils or calyxes are little white hairlike designs that will continue to develop and at last form into ragged blossoms.

Males can be distinguished in light of the fact that as opposed to developing pistils the dust sacs (little green balls) will remain 'unsprouted' and you will not have the option to detect any pistils.

You can likewise decide to dispose of male plants yet this is absolutely dependent upon you. Male plants are generally discarded by experts since it builds the gamble of your females being pollinated by the guys.

which may result in female plants that are weaker and have less powerful flower buds. Additionally, you will be spending money and time on a plant that will not produce any smoky flowers.

## TIME TO HARVEST (PICK AND TRIM)

The trick here is understanding when is the best chance to reap your bloom buds. The collection window will begin when your blossoms quit developing white hairs. From that point forward you'll have three choices to chop down your bloom buds:

Around half of the white hairs have turned hazier for low potencies.

At the point when somewhat more than half of the hairs have changed for the greatest THC intensity.

At the point when most hairs have turned more obscure for powerful loosening up impacts however lower THC levels.

Picking and managing are not difficult to do in principle. You simply have to cut your blossom from the plant with a scissor and afterward trim the buds from their stems.

## DRY AND CURE FLOWER BUDS

Restoring and drying your buds is the step where you put the cherry on top of this long cycle and pass on your blossom prepared to smoke.

Take your picked and managed buds and hang them topsy turvy in a cool dull spot. A wardrobe with enough ventilation and low moisture could do. Simply attempt to avoid kitchens or washrooms where dampness will in general be higher.

You'll realize the buds are dry enough when the more slender stems snap and the thick stem is bendy.

You can now remove these, trim the stems to the greatest extent possible, and place them in glass mason jars. The containers ought to be firmly fixed and completely filled.

Open the containers consistently for a couple of moments for the initial 3 weeks of the restoring system. Contact your buds and feel in the event that they are holding dampness or not. On the off chance that you really can't dispose of dampness, you could likewise utilize dampness packs that assist you with controlling the dampness.

After buds have felt dry for an entire week each time you open the containers to mind them you'll have the option to scale back to open the containers one time per week. Organizing can require as long as 30 days however time will rely upon you and your own taste. Toward the day's end, you're developing your plants for you and not for any other person.

# HOW TO GROW WEED AT HOME: THINGS TO LOOK OUT FOR

Like with different plants, you'll need to screen your weed plants routinely and guarantee they're becoming true to form.

Some the things to keep an eye out for are:

Change coloring or spotting

Leaves tumbling off, twisting up, or dying

Smell

Very slow growth

Stretching

Bugs

## THINGS TO KNOW ABOUT INDOOR GROWING

Indoor development is a shockingly modest choice. Without a doubt, you'll require underlying speculation forthright, however, it's not quite as costly as you would have first thought. Particularly on the off chance that you're going for a little development operation of only a couple of plants.

You could try and have the option to get imaginative and fabricate your development framework yourself. Simply ensure it has adequate room for air and no extreme measures of moistness and temperatures.

Inside development operations give a simpler method for controlling your 'fixing' amounts or the admittance to different factors that your plants are getting. Something you won't have the option to do on outside harvests.

While indoor developed plants can create buds all the more reliably, they are likewise more dependable on you. So they will request a greater amount of your time than an outside developed blossom.

**Normal temperatures to follow:**
**Young plants**: 20-30 C.
**Flowering Stage:** 18/26 C.
In encased regions, various frills like fans and exhaust frameworks may be required. As well as controlling the temperature radiated by your lighting framework.

# BENEFITS OF INDOOR CANNABIS GROWING

In contrast to outdoor cultivation, indoor cannabis growing offers a number of advantages, including the following: Safety: Even in states where marijuana is legal, judgmental neighbors—as well as thieving neighbors—may still be around, waiting to point the finger at your "hobby." Developing weed inside could assist with keeping away from only that.

**You're in charge of blossoming**: You can "tell" your plants to blossom at whatever point you need, by simply changing the light cycle. This also means that depending on how frequently you start the flowering phase, you can have multiple harvests.

**You can establish it all year**: Whether you believe that you should do it all through summer or winter, you can develop your weed plants nonstop, bother-free. You're not attached to specific seasons like individuals who favor developing outside are.

**Quality buds**: While indoor development may be costlier, you can tailor your development climate to deliver buds of top caliber. You just need to change the provisions (lighting, supplements, develop medium) appropriately.

There are several indoor-friendly strains to choose from: Favorite banks stock an entire heap of indoor-accommodating strains, for your pick.

One good example is ILGM, which has a lot of indoor strains, including the ones listed below:

**Blue Dream**
**White Widow**
**Gorilla Glue**
**Bergman's Gold Leaf**
**Young Lady Scout Treats Outrageous**

## THINGS TO KNOW ABOUT OUTDOOR GROWING

Whenever people first ran into weed they ran into open-air development landraces that had been partaking in an untamed life for some time. We don't have the foggiest idea who the virtuoso that proposed smoking it was, however, the fact of the matter is that they were developing outside.

Therefore, outdoor growth operations are effective. You simply have to track down the right environment. Not over the top humid or dry, not excessively cold or hot. Not a simple accomplishment, but rather regardless of whether you have the ideal temperature there's nothing that changing watering amounts can't assist you with.

Open-air development is a lot less expensive, predominantly in light of the fact that the sun is doing the truly difficult work for you. Meaning no underlying interest in lighting frills or a higher energy bill.

Notwithstanding, open-air development is less dependable, less private, and carries more risk to your plants.

## ADVANTAGES OF GROWING MARIJUANA OUTDOORS

Obviously, developing Maryjane outside can likewise be fascinating on the off chance that you have a major nursery and you're not stressed over your neighbors.

## The following are a couple of advantages of open-air developing

**Low expenses** - Since you're involving the sun as your principal wellspring of energy, you don't need to pay for indoor lights and light clocks.

All the more critically, you won't require an air conditioner unit or a dehumidifier.

**Your plants can survive mistakes**- While developing inside, you can undoubtedly kill your plants on the off chance that the room gets excessively hot or excessively moist, which happens to numerous novices. With outside development, your plants don't require a lot of support and not much can turn out badly.

**Greater yields -** While indoor yields are by and large restricted, open-air plants can develop as large as you need them to. Thus, this implies it's normal for certain strains to yield up to 1000 g for each plant.

**Perfect for Sativas**: As you are aware, Sativa plants can reach 10 feet in height. This makes outside development ideal in the event that you don't have high roofs.

## HOW TO GROW CANNABIS: YOUR QUESTION ANSWERED

### How do I control weed odor?

Indoor growers will probably experience weed odor, particularly during the flowering stage, whereas outdoor growers will probably not. Be that as it may, stress not, this is the way to go about it.

### Use carbon channels (also known as carbon scrubbers):

Place these at the highest possible height in your grow room.

### Scent-engrossing gels:

While these probably won't dispose of the weed smell, they'll veil it, supplanting it with different fragrances.

## Make sure the air flows properly:

This assists control moistness and temperature, assisting with eliminating weed smell. Fans might prove to be useful here.

## Really look at the degrees of temperature and dampness:

High temperature (and mugginess) will probably raise the marijuana smell. You should furnish your developing region with a humidifier to assist with cutting it down.

## How long does it take for cannabis to grow indoors?

Developing weed inside can take anyplace between 20 to 30 weeks or more. When you have an indoor development room arrangement, an excellent seed will grow in 3 to 10 days. From here, it'll require nearly 2 to 3 weeks to develop into a seedling (otherwise known as the seedling stage). The plant will then enter the vegetative stage, which can last anywhere from three to sixteen weeks after it has been transplanted to a new growing environment. Your plants ought to develop rapidly at this stage as they ingest increasingly more carbon dioxide and supplements. Then, your pot plant(s) will change to the blooming stage, which will probably last 8 to 11 weeks.

Keep in mind, these development periods are simply gauges. While developing inside, there are different circumstances (counting strain type) that can influence the time it takes to develop (and reap) your weed plants.

## What are the factors to consider before growing weed?

**Budget**: Try not to stall and get out. Set a budget that you can stick to, especially if you grow indoors. Take into consideration the prices of supplies, lighting, growing space, fans, and other things.

**Grow space**: You need to begin little, particularly on the off chance that you're a first-time producer. Make sure your grow area has water and air inlets and outlets as well as the right temperature.

**Privacy**: Is your developed region private? That's just so that some neighbors, who judge each other, can have their "peace." You get the point.

**Yield**: You might want to select marijuana strains with high yields if you want a high yield. In the interim, on the off chance that you wouldn't fret this, you can snatch pretty much any feminized (or auto-flowering) stuff, given it's top notch.

**Time**: Given that the majority of marijuana strains mature in at least three months, *how much time are you willing to devote to growing it?* Therefore, devote some time each week to tending to your cannabis plants.

# CONCLUSION

## HOW DO YOU GROW WEED - A RECAP

Developing Maryjane need not be muddledYou can save money by growing your own marijuana instead of purchasing it grown. You just need to do a certain thing: follow the fundamental advances strictly, particularly if you need to develop pot inside.

Ensure you're utilizing great seed strains, as well. We trust this guide assists you with establishing your own weed problem free — so you disregard the issue of purchasing the thing. Because the seed bank sells high-quality strains that are guaranteed for germination, experienced growers will tell you that ILGM is the best place to buy seeds.

They likewise bundle stuff prudently and transport these subtly (so your meddling neighbors don't have to be aware). Furthermore, indeed, they couldn't be more creative with regard to such things as development guides. They have a supportive client work area, as well, notwithstanding an educational blog and gathering. Developing pot need not be convoluted, given you stick to the means in this aide. Best of luck!

# CHAPTER 2:

## 7 COMMON CANNABIS GROWING MISTAKE IN THE INDOOR GROW ROOM AND HOW TO FIX IT

When you first start growing cannabis, you are likely to make some mistakes. Only one out of every odd gathering will emerge true to form, yet careful discipline brings about promising results. The more cycles you go through, the better you'll grasp the plant and the effect of the developing climate.

But why enter the procedure with no knowledge? A little research will help you avoid some of the most common mistakes, whether you are just setting up your grow room or working on your next crop.

From overwatering to overloading to the drying system, the following are seven of the main issues home producers face.

# 1.Choosing the wrong genetics

The strain you decide to develop can represent the moment of truth you reap. Business cultivators frequently evaluate various strains under different natural circumstances to reveal the best outcomes. However, as a novice indoor grower, you do not have this flexibility.

To get the hereditary qualities right all along, you'll need to search for cultivars with fitting attributes for your current circumstance. You'll likewise need to source seeds or clones from confided-in sources to guarantee their practicality and improve the probability of getting a top-notch item.

## 2. Overwatering

One more typical issue is the propensity to overwater. Albeit numerous new cultivators accept the dirt ought to continuously feel soaked, this much water will begin to suffocate the roots and hinder improvement. In the worst situation imaginable, it might kill your plants.

Weed roots, similar to the leaves, ingest oxygen. In the event that you are watering excessively or too habitually, they won't be able to dry out and, in this way, can't relax. If the top one to two inches of the substrate have dried out, it's usually time to water again.

## 3. Overfeeding

In addition to overwatering, many novice growers feed their plants too much. In the event that you add an excessive number of supplements too regularly, you're squandering costly added substances, yet risk artificially consuming your plants. This is particularly normal with novices stringently following the excessively forceful taking care of timetable suggested on the rear of supplement holders.

With these high nutrient levels, experienced growers may be able to balance plant stress and environmental inputs, but novice growers may not always be able to manage it. In this way, assuming you notice indications of leaf consumption, take a stab at decreasing the sum as well as the recurrence of supplement application.

## 4. Overlooking pH Level

pH level is a basic part of the indoor developing climate, however, one which beginner cultivators oftentimes overlook. Assuming the pH level of your supplement arrangement and substrate is excessively high (over 7), your plants will encounter a supplement lock-out.

On the off chance that the pH level is excessively low (under 5), there is a gamble of expanded supplement accessibility, explicitly iron and manganese. An abundance of supplement retention can prompt supplement harmfulness. Thus, it's enthusiastically prescribed to put resources into a pH testing unit, regularly test your run-off, and go for the gold between 5.8 to 6.2.

## 5. Lighting is inefficient and ineffective.

In the past, everyone used to grow cannabis indoors using powerful HID lighting. While these installations created a light power ideal for pot, they ran unbelievably hot and were a long way from energy production.

Fortunately, today you can get a similar force from cutting-edge Drove develop lights however with a much lower power bill and less generally speaking intensity creation. Besides, many new plans are unequivocally designed to create the light range and power that pot plants need to flourish. LEDs are therefore a safer and more effective choice, particularly for home growers.

## 6. Wrong Harvest Time

Getting harvest time right decides the plant's last phytochemical profile and power. Assuming you collect too soon, you will miss top strength; too late, and the molecules of THC begin to break down.

In the final week of flowering, color the trichomes on a daily basis to achieve maximum potency. Clear and clear trichome heads show that it's not exactly time to collect, while golden shading demonstrates you've missed the best window. When the majority of the trichomes are cloudy white, aim to harvest.

A jeweler's loupe or magnifying glass can help you see the color and transparency of tiny trichome heads, which are unnoticeable to the naked eye.

## 7. Risky Drying Environment

Keep in mind, developing pot doesn't stop at the slash. Pot requires the same amount of consideration during the drying and restoring stages as it does all through its developing cycles.

The post-reap climate influences the cannabinoid and terpene profile nearly as much as the most recent couple of long stretches of bloom. Encouraging the growth of mold, mildew, and other microorganisms, can also affect the integrity of the flower if the conditions aren't right.

After you gather your plants, keep them in a dull room with a legitimate, continuous wind stream. Utilize a natural control to keep up with the room between 65 to 75 °F with 45 to 55 percent stickiness.

This climate saves the phytochemical profile while decreasing the gamble of post-reap bud decay.

## 8. Not Keeping It Private

The most crucial point to keep in mind is not to tell anyone about your expanding profession. During the few months of the developing undertaking, the most effective way to safeguard it is to simply keep it as hidden as possible - the ideal choice will be for you to be the only one knowing.

Every person you tell increases your risk of being robbed or, even worse, of being punished legally. Consider this when you are getting the developing space itself - this is your space, so pick a confidential one. This applies even where it is legitimate to develop marijuana. The less individuals know, the better.

## 9. Using the wrong fertilizer

While picking a compost, you want to focus on the NPK proportion to have a decent reap toward the end.

During every development cycle, you really want to utilize a particular manure that has the perfect proportion of nitrogen, which the plants need, particularly during the blooming cycle.

## 10. Over Pruning

Pruning is likewise fundamental for helping plant development at the same time, comparatively to watering and taking care of, over pruning is another normal slip-up. Pruning the plants a lot can prompt debilitating the plant and even kill them. You must be mindful so as not to prune the entire plant but rather to keep up with the right strategy and consistently prune short of what you figure you could need to.

## 11. Not having the right protection (or any!)

Working any business is hazardous and developing yields is no special case. In the event of a loss—from fire, theft, wind, etc.—covering your product can provide financial assistance. but can also aid in time protection.

This developing process requires commitment and exertion, losing the work you have given a long time to can be baffling, however with the right protection inclusion the most common way of beginning new will be more straightforward.

However, self-protection may be even more crucial. The guideline is weighty in the marijuana space and with a Chiefs and Officers insurance contract you can safeguard your organization's resources as well as your very own resources in the event that you disregard guidelines and are fined/punished. There are eyes on the weed business so ensure you are not seriously endangering yourself while plunging into it!